LESIONS OF BREAST – A REVIEW

Dr. Gunvanti Rathod

Dr. Pragnesh Parmar

Dr. Sangita Rathod

Dr. Ashish Parikh

LESIONS OF BREAST – A REVIEW

Dr. Gunvanti Rathod, MD (Pathology)

Assistant Professor
Department of Pathology
SBKS Medical Institute and Research Centre
Vadodara, Gujarat, India

Dr. Pragnesh Parmar, MD (Forensic Medicine)

Assistant Professor
Department of Forensic Medicine
SBKS Medical Institute and Research Centre
Vadodara, Gujarat, India

Dr. Sangita Rathod, MD (Medicine)

Assistant Professor
Department of Medicine
AMCMET Medical College
Ahmedabad, Gujarat, India

Dr. Ashish Parikh, MD (Medicine)

Consultant Physician
Gayatri Hospital
Gandhinagar, Gujarat, India

DEDICATION

This book is dedicated to my loving daughter **Jayani**.

- **Dr. Gunvanti Rathod**

ACKNOWLEDGEMENTS

We acknowledge the immense help received from the scholars whose articles are cited and included in references of this book. The authors are also grateful to authors / editors / publishers of all those articles, journals and books from where the literature for this book has been reviewed and discussed.

We express our gratitude to our parents and in-laws for their constant encouragement, support and blessings.

It will be an injustice if we do not thank all our students for their innovative ideas and feedback.

CONTENTS

TOPIC	PAGE NUMBER
INTRODUCTION	7
LABORATORY DIAGNOSTIC PROCEDURE FOR BREAST LESIONS	8
FNAC OF BREAST	9
CORE NEEDLE BIOPSY IN BREAST LESIONS	13
HISTO-PATHOLOGICAL EXAMINATION OF WHOLE SPECIMEN	14
HISTOLOGY AND PHYSIOLOGY OF BREAST	15
VARIOUS BREAST LESIONS	18
CYTOLOGICAL AND HISTOLOGICAL FINDINGS OF VARIOUS BREAST LEASIONS	22
REFERENCES	46

INTRODUCTION

- The incidence of breast cancer is slowly increasing and an estimated 17,50,000 were diagnosed in 1991. [1] At present, it will strike 1 in 9 women over a lifetime. Breast carcinoma is known to occur in older age group especially peri-menopausal and post-menopausal. The importance of early diagnosis cannot be underscored as management and survival is markedly improved with early stage of breast cancer. Nowadays younger patients are being diagnosed with breast cancer. Early stage diagnosis is even more important here, so that an option of breast conservation can be offered.

LABORATORY DIAGNOSTIC PROCEDURES FOR BREAST LESIONS

1. FNAC (Fine needle aspiration cytology)

2. Core needle biopsy examination

3. Histopathological examination of whole specimen

FNAC OF BREAST

- FNAC (Fine Needle Aspiration Cytology) can be performed on any clinically palpable lumps or roentgenographically detected non-palpable lesions of the breast. FNAC also has very few complications.

HISTORICAL PERSPECTIVES

- Fine needle aspiration cytology is the sampling of a palpable or roentgenographic mass by means of fine needle with negative pressure applied by attached syringe.
- The concept of diagnostic cytology has been greatly influenced by the work of **G. N. Papanicolou** in 1941. [10] His contribution for cytologic presentation of cancer has changed the status of cytology from a largely theoretical field of knowledge to a widely acceptable laboratory procedure.
- **Martin and Ellis** (1930) were the first to use aspiration technique for obtaining diagnostic material for cytologic study. [11]
- **Dudgeon** in 1934 [12] claimed that aspiration cytology has definite advantages over section study.
- **Gupta et al.** in India (1975) studied total 222 cases by fine needle aspiration cytology in various primaries and secondaries accessible tumors and overall positive results found in 84.8 % with one false negative, one was suspicious and two were inconclusive . [13]
- **O'neil et al.** (1997) reported a sensitivity of 97 %, specificity of 78 %. Overall positive and negative predictive values were both 92 %. The accuracy of FNAC in the series was 92 %. [14]
- **Hunt C M et al.** in 1990 made a successful attempt towards the grading of breast cancers from cytological criteria alone. [15] Thus sensitivity of the FNAC is very high and in developing countries where the patient load is far higher than the available resources, a technique like FNAC is very effective for early diagnosis and triage of patients so that maximum benefits can be given to the patients at the earliest with the available resources. It provides a

simple and rapid diagnostic method for breast lesions. [2] Of course, it serves as a complement and not a substitute to the histopathological diagnosis. [15]

INDICATIONS OF FNAC BREAST: [3, 4]

The use of FNAC can be defined according to an increasing scale to suit local experience and requirements. Most important of them are:

1. The preoperative confirmation of clinically suspected cancer.
2. The investigation of clinically palpable, clinically benign or malignant, as a guide to clinical management.
3. The investigation of suspected recurrence or metastasis in cases of previously diagnosed cancer.
4. The confirmation of inoperable locally advanced cancer.
5. As a complement to mammography in screening situation.
6. The diagnosis of simple cysts.

COMPLICATIONS: [4]

- Major hematomas are unusual. Pneumothorax is rare but an important complication.

- A chance of tumor seeding along the needle tract is more of a theoretical proposition awaiting documentary evidence. There is no influence on long term survival rates from FNA procedure per se.

LIMITATIONS OF FNAC: [3, 4]

1. FNA cytology requires training in the preparation of quality smears and considerable cytology expertise is required to interpret FNA cytology.
2. Cytological examination on its own cannot decide, invasion is present or not.

3. Definite diagnosis of some lesions can be difficult to make on the basis of cytology. These include: Low grade DCIS, atypical ductal hyperplasia, tubular carcinoma etc.
4. FNA cytology may not be the sampling technique of choice for lesions that are relatively hypocelluler and yield scanty epithelial material. These include sclerotic fibroadenomas, sclerosing ductal carcinoma, and infiltrating lobular carcinoma.

Conditions in which there is risk of false positive diagnosis: [5]

1. Papillary lesions
2. Epithelial hyperplasia with nuclear atypia
3. Fibroadenoma with atypia
4. Pregnancy and lactation
5. Skin adnexal tumors

Conditions in which there is risk of false negative diagnosis:

- A false negative diagnosis is most often due to sampling problems; interpretation errors are rare. [6] However, tumor size, paucicellularity and special type histologic factors can contribute to an increase in the false negative rate [7] namely.
  - ➢ Low grade ductal carcinoma, tubular carcinoma.
  - ➢ Lobular carcinoma.
  - ➢ Tumors with central necrosis or sclerosis.
  - ➢ Small carcinoma next to a dominant benign lesion.

TRIPLE TEST:

- FNA is never used as the sole diagnostic modality in determining intervention for palpable breast lesions. The triple test, a combination of physical examination, mammography and FNA has been shown to be highly sensitive and specific in the diagnosis of cancer. The triple test is positive, and negative if all the components are negative. Of the three tests, FNA has

11

the highest sensitivitiy and specificity and may prompt further action in cases in which physical examination and mammography are not suspicious.

CORE NEEDLE BIOPSY IN BREAST LESIONS

- Core biopsy yields tissue fragments allowing architectural features of the lesion to be identified to determine whether DCIS or invasive carcinoma is present. Core biopsy is useful in the evaluation of lesions likely to be low histopathological grade and in those presenting as architectural distortions, for which FNA cytology may fail or has failed to prove a diagnosis, Core needle biopsy does provide higher specificity.
- FNA has been shown to be more sensitive than core biopsy. [8]
- Although core needle biopsy provides a larger volume of breast because of the technique by which the sample is obtained. Core biopsy is definitely a robust and reliable diagnostic modality, but carries with it disadvantages in terms of a longer turn- around due to the tissue processing time, patient discomfort during the procedure [9] and higher complication rate.

HISTOPATHOLOGICAL EXAMINATION OF WHOLE SPECIMEN

- Histopathological examination is the key to diagnose the type of breast lesion. Histopathological examination of breast reduction specimens may reveal important pathological diagnoses. In some cases, patients were discharged from medical care despite histopathological examination revealing lesions associated with an increased risk of developing breast carcinoma.

- One investigation highlighted as making little contribution to patient care was the histopathological examination. The document stated that sections from macroscopically abnormal areas were justified, but that the value of random histology appeared minimal. There is little evidence to support or refute this statement and a college audit was recommended.

- The finding of important pathological diagnoses in macroscopically normal breast tissue raises the possibility that important diagnoses are missed as a result of sampling errors. These specimens are often very large and the selection of two blocks of tissue for microscopic analysis means that only a very small proportion of the total tissue submitted is examined histologically. Detailed macroscopic examination with palpation of the tissue is used to identify areas that may contain microscopic abnormalities. However, unless the entire specimen is processed for microscopy, with the serious cost and time issues that this would entail, this problem cannot be completely overcome.

HISTOLOGY AND PHYSIOLOGY OF BREAST

HISTOLOGY OF BREAST: [1, 2]

The breast of adult, non pregnant female in a resting phase has the following features:

- The morphological unit of the organ is the single gland, a complex branching structure that is composed of two major parts: the terminal duct lobular unit (TDLU) and the large duct system.

- The TDLU is formed by the lobule and terminal ductule and represents the secretory portion of the gland. It connects with the subsegmental duct, which in turn leads to segmental duct and this to a collecting duct which empties into the nipple.

- A fusiform dilatation located beneath the nipple is called as the lactifereous sinus.

- The TDLU is recognized by its distinctly lobular architecture; presence of a mantle of a specialized, myxoid appearing hormone responsive connective tissue; and the absence of elastic fibres. The larger ducts have a lesser amount of specialized stroma and are enveloped by a continuous and well developed layer of elastic tissue.

- The entire ductal – lobular epithelial system of the breast is covered by a specialized two cell type epithelial lining: the inner epithelium with secretory and absorptive functions and the outer myoepithelial cells.

- The entire glandular system rests on a continuous basement membrane.

- The nipple has a characteristic microscopic appearance. In addition to the large collecting ducts opening into the surface, it contains numerous

15

sebaceous glands that open independently of hair follicles and a dense fibrous stroma in which erectile smooth muscle tissue is embedded. Montgomery's tubercles are areolar protuberances, usually 10 to 20 in number, which become prominent during pregnancy; microscopically, they are formed by a collecting duct associated with a sebaceous apparatus. The epidermis of the nipple and areola resembles that of the skin elsewhere, except for an increase in melanin content in the basal layer and the occasional presence of basally located clear cells known as Toker's cells, which are related histogenetically to Paget' s disease.

PHYSIOLOGICAL CHANGES IN MAMMARY BREAST: [2]

- Breast tissue responds markedly to hormonal and other influences. It may display a wide range of normal appearances: the immature and largely resting breast before puberty; the developed breast of reproductive life, the actively secreting breast of lactation and the involuted post menopausal breast.

- **Puberty breasts:** They resembles gynecomastia, because it is characterized by epithelial hyperplasia in duct system with a solid and micro papillary configuration, growth and branching of proliferating ducts. There is increase in number of ducts cross sections. Prepubertal enlargement is due to increased fat and connective tissue.

- **Pregnancy:** Lobular growth is due to cellular proliferation. Myoepithelial cells continue to remain evident in ducts but are obscured by expanded lobules. During pregnancy and lactation, all cell types show a high level of proliferative activity. During lactation, cytoplasm is vacuolated and secretion is accumulated. Involution of breast, after lactation ceases, occurs over a period of three months.

- **Menopause:** There is decrease in cellularity and number of lobules with thickening of lobular basement membrane and collagenization of stroma.

16

Atrophy tends to spare myoepithelial cells. Clear cell changes occur in both pre and post menopausal women.

VARIOUS BREAST LESIONS

WORLD HEALTH ORGANIZATION (WHO) 2003 CLASSIFICATION [3]

1. **Epithelial tumors**
   - 1.1 Invasive ductal carcinoma
   - 1.2 Invasive lobular carcinoma
   - 1.3 Tubular carcinoma
   - 1.4 Invasive cribriform carcinoma
   - 1.5 Medullary carcinoma
   - 1.6 Mucinous carcinoma and other tumors with abundant mucin
     - 1.6.1 Mucinous carcinoma
     - 1.6.2 Cystadenocarcinoma and Columnar cell mucinous Carcinoma
     - 1.6.3 Signet ring
   - 1.7 Neuroendocrine tumors
     - 1.7.1 Solid neuroendocrine carcinoma
     - 1.7.2 Atypical carcinoid
     - 1.7.3 Small cell/ oat cell carcinoma
     - 1.7.4 Large cell neuroendocrine carcinoma
   - 1.8 Invasive papillary carcinoma
   - 1.9 Invasive micropapillary carcinoma
   - 1.10 Apocrine carcinoma
   - 1.11 Metaplasic carcinoma
     - 1.11.1 Pure epithelial metaplasia
       - 1.11.1.1 Squamous cell carcinoma
       - 1.11.1.2 Adenocarcinoma with spindle cell metaplasia
       - 1.11.1.3 Adenosquamous carcinoma
       - 1.11.1.4 Mucoepidermoid carcinoma
     - 1.11.2 Mixed epithelial/ mesenchymal metaplastic carcinoma
   - 1.12 Lipid rich carcinoma
   - 1.13 Secretory carcinoma
   - 1.14 Oncocytic carcinoma
   - 1.15 Adenoid cystic carcinoma

1.16   Acinic cell carcinoma

1.17   Glycogen rich clear cell carcinoma

1.18   Sebaceous carcinoma

1.19   Inflammatory carcinoma

1.20   Lobular neoplasia

         1.20.1 Lobular carcinoma in situ

1.21   Intraductal proliferative lesions

         1.21.1 Usual ductal hyperplasia

         1.21.2 Flat Epithelial atypia

         1.21.3 Atypical ductal hyperplasia

         1.21.4 Ductal carcinoma in situ

1.22   Microinvasive carcinoma

1.23   Intraductal papillary neoplasms

         1.23.1 Central papilloma

         1.23.2 Peripheral papilloma

     1.23.3 Atypical papilloma

     1.23.4 Intraductal papillary carcinoma

     1.23.5 Intracystic papillary carcinoma

1.24 Benign epithelial proliferations

     1.24.1 Adenosis including variants

     1.24.1.1 Sclerosing adenosis

     1.24.1.2 Apocrine adenosis

         1.24.1.3 Blunt duct adenosis

         1.24.1.4 Microglandular adenosis

1.25 Adenomas

         1.25.1 Tubular adenoma

         1.25.2 Lactating adenoma

         1.25.3 Apocrine adenoma

         1.25.4 Ductal adenoma

2. Myoepithelial lesions

2.1   Myoepitheliosis

2.2   Adenomyoepithelial adenosis

2.3   Adenomyoepithelioma

2.4   Malignant myoepithelioma

3. Mesenchymal lesions
    3.1   Hemangioma
    3.2   Angiomatosis
    3.3   Hemangiopericytoma
    3.4   Pseudoangiomatous stromal hyperplasia
    3.5   Myofibroblastoma
    3.6   Fibromatosis
    3.7   Inflammatory myofibroblasic tumor
    3.8   Lipoma
                3.8.1 Angiolipoma
    3.9   Granular cell tumor
    3.10   Neurofibroma
    3.11   Schwannoma
    3.12   Angiosarcoma
    3.13   Liposarcoma
    3.14   Rhabdomyosarcoma
    3.15   Osteosarcoma
    3.16   Leiomyoma
    3.17   Leiomyosarcoma

4. Fibroepithelial Tumors
    4.1   Fibroadenoma
    4.2   Phylloides tumor
            4.2.1 Benign
            4.2.2 Borderline
            4.2.3 Malignant
    4.3   Periductal stromal sarcoma, low grade
    4.4   Mammary hamartoma

5. Tumors of the nipple
    5.1   Nipple adenoma
    5.2   Syringomatous adenoma

5.3   Paget disease of the nipple

6. Malignant lymphoma
6.1 Diffuse large B - cell lymphoma
6.2   Burkitt lymphoma
6.3   Extranodal marginal – zone B- cell lymphoma of MALT type
6.4   Follicular lymphoma

7. Metastatic tumors

CYTOLOGICAL AND HISTOLOGICAL FINDINGS OF VARIOUS BREAST LESIONS

CYTOLOGICAL AND HISTOLOGICAL FINDINGS OF VARIOUS BREAST LESIONS [4, 1]

COMPONENTS OF NORMAL BREAST TISSUE:

CYTOLOGY:
- Overall low cell yield.
- Sheets of ductal cells and aggregates of ductal epithelial cells with small uniform nuclei.
- Myoepithelial nuclei visible among epithelial cells in aggregates.
- Single, bare, oval nuclei separate from epithelial aggregates.

LACTATION AND PREGNANCY:
- Aspiration is often milky.
- Duct epithelial cells are present in sheets, branching groups or synctitial clusters.
- Foamy macrophages are seen.
- Enlarged nuclei with single prominent nucleoli are often identified.
- FNAC in pregnancy is difficult to interpret unless history of pregnancy is provided.

FAT NECROSIS:
- A background of granular debris, fat and fragments of adipose tissue.
- Foamy macrophages, multinucleated giant cells and adipocytes with bubbly cytoplasm.
- Chronic inflammatory cells.
- Absence of epithelial cells.

ORGANIZING HEMATOMA:
- Aspirate is thick and hemorrhagic with dark brown blood.
- Cholesterol crystals along with pigment containing macrophages.
- Plump fibroblasts are in large numbers.

MASTITIS:
- A benign bimodal pattern.
- Inflammatory cells chronic and/or acute.
- Regenerative epithelial atypia.
- Epithelial histiocytes, multinucleated giant cells and many plasma cells (Granulomatous mastitis)

BENIGN EPITHELIAL HYPERPLASIA, ADENOSIS etc.:
- A bimodal population of sheets and clusters of epithelial cells and single bare nuclei.
- Cyst macrophages and / or epithelial cells of oxyphil / apocrine change.
- Mild epithelial atypia (variable).

HISTOLOGY:

ADENOSIS:
It applied to any hyperplastic process that involves the glandular component of the breast.
  A. **Blunt duct adenosis.**
  B. **Sclerosing duct adenosis:** Nodule retains a round configuration and is more cellular centrally than peripherally. The elongated and compressed proliferating tubules are lined by two cell types. Pleomorphism and necrosis are absent.
  C. **Microglandular adenosis:** Small uniform glands are with open lumina containing eosinophilic secretion are distributed in an irregular fashion within fibrous tissue. The glands are lined by a single layer of cells; the myoepithelial layer may be absent.

FIBROCYSTIC CHANGES:

- Cyst macrophages and sheets of duct epithelial cells of oxyphil or apocrine type are found. Macrophages have an abundant, finally vacuolated cytoplasm which may contain pigment and small round central nuclei.

- **HISTOLOGY:** Basic morphological changes include cyst formation, apocrine metaplasia, fibrosis, calcification, chronic inflammation, epithelial hyperplasia and fibroadenomatoid changes.

FIBROADENOMA:

- A high cell yield.
- Large branching, monolayer sheets of uniform epithelial cells.
- Numerous single, bare nuclei of benign type.
- Fragments of fibromyxoid stroma.

- **Gross:** Sharply demarcated, firm mass. The cut surface is solid, grayish white, bulging with a whorl – like pattern and slit like spaces. Necrosis is absent.

- **Microscopically:** Tubules are composed of cuboidal or low columnar cells with round uniform nuclei resting on a myoepithelial cell layer. The stroma is usually made of loose connective tissue rich in acid mucopolysaccharides. Elastic tissue is absent.

PHYLLOIDES TUMOR:

- **CYTOLOGY:** Predominance of stromal over epithelial components. Numerous single bare nuclei in the background, many of which are spindle shaped rather than fibroblastic.

- **HISTOLOGY:** The two key features of phylloides are stromal hypercellularity and presence of benign glandular elements as an integral component of the neoplasm.

- **Low grade malignant or borderline phylloides:** Mitoses 2-5/10 hpf, increased cellularity, and invasive borders.

- **Malignant phylloides:** Mitoses > 5-10 hpf, invasive borders and stromal cellular pleomorphism.

DUCTAL CARCINOMA IN SITU:

Large cell type (comedocarcinoma)
- Neoplastic cells in irregular aggregates and single.
- Large pleomorphic cells showing obvious malignant features.
- Necrotic debris, lymphocytes and vacuolated macrophages.

- **HISTOLOGY:** The ducts show a solid growth of large pleomorphic tumor cells accompanied by generally abundant mitotic activity and lacking connective tissue support. Necrosis is always present and constitutes an important diagnostic sign. Coarse calcification may be present, myoepithelial cells are usually absent and stroma around the involved duct shows concentric fibrosis.

Papillary type
- Papillary aggregates, sometimes with a central fibrovascular core.
- Columnar cells in row palisades and single.
- Variable nuclear enlargement, pleomorphism , and atypia.
- Macrophages and epithelial cells with cytoplasmic vacuoles.
- Necrotic debris.

- **HISTOLOGY:** Features favoring malignancy are uniformly in size and shape of epithelial cells, presence of one cell type only, nuclear hyperchromasia, high N: C ratio, high mitotic activity, scanty or absent stroma.

25

Cribriform type

- Epithelial cells relatively cohesive forming large monolayer, balls or papillary fragments.
- Macrophages, often with hemosiderin pigment.

- **HISTOLOGY:** Round regular spaces are formed within glands, the more regular these spaces are in size, shape and distribution the more likely the lesion is malignant.

LOBULAR CARCINOMA IN SITU:

- The lobules are distended and completely filled by relatively uniform, round, small to medium sized cells with round, normochromatic nuclei. Atypia, pleomorphism, mitotic activity, and necrosis are minimal or absent, and there is some lack of cohesiveness.

MALIGNANT EPITHELIAL TUMORS:

- Breast cancer is the most common cancer in women; it accounts for almost one quarter of all cancer in women and is more common in western affluent countries. [5] Despite its increasing incidence throughout the world, breast cancer is still regarded as predominantly a disease of postmenopausal women. The disease is uncommon in women aged less than 30. The National Cancer Institute's surveillance, Epidemiology and End results database reported that out of 77, 368 women diagnosed with breast cancer only 1 % were between the ages of 20 and 29. [6]
- It has been widely reported in the literature that breast cancer diagnosed in younger women tends to be more advanced with poor prognosis. [7, 8, 9, 10, 11, 12]
- **Fernandopulle et al.** [13] has reported in increasing proportion of young breast cancers i.e in women less than 35 years of age.
- The prognosis of breast cancer is very good if detected at an early stage, the survival has significantly improved over the last three decades and more so since 1990s mainly due to early detection by screening mammography and advances made in adjuvant chemo and hormonal therapy modalities. In

addition to the stage, histologic type, grade and hormonal status of the tumor are important prognostic factors.

- Two main histologic groups of mammary carcinoma include ductal and lobular carcinoma both with their in situ and invasive counterparts and both with differing tumor biology and clinical manifestation. Most mammary carcinoma are thought to arise from the terminal duct lobular unit (TDLU), which may then differentiate into a ductal or lobular phenotype. Within the invasive mammary carcinoma group, invasive ductal carcinoma is the most common type of breast malignancy accounting for 65-80 % of all mammary carcinomas. [14] Invasive ductal carcinoma is a heterogonous group of tumors, majority of which comprise invasive ductal carcinoma NOS (Not Otherwise Specified) or NST (No Special Type) or usual type. These are tumors that do not exhibit sufficient histologic characteristics to be included in to special type of invasive ductal carcinoma such as tubular, medullary, cribriform, micropapillary etc.

INFILTRATING DUCT CARCINOMA - NOT OTHERWISE SPECIFIC (NOS):

- A high cell yield.
- Irregular, angulated, clusters of atypical cells.
- Nuclear enlargement and irregularity of variable degree.
- Single cells with intact cytoplasm.
- Necrosis important if present.

- **HISTOLOGY:** Classic (NOS) invasive ductal carcinoma – tumor grows in diffuse sheets, well defined nests, cords or individual cells. Glandular differentiation may be well developed or altogether absent. The tumor cells vary in size and shape, their nuclei and nucleoli are prominent, mitotic figures are numerous.

VARIANTS OF INVASIVE CACINOMA:

- Special types of invasive mammary carcinoma are defined by specific histologic criteria that recognize a clustering of features in some breast cancers. Through rigorous application of histopathological definitions, subsets of invasive carcinoma maybe recognized and usually indicate an excellent prognosis.

Series	NST	Lobular	Medullary	Tubular	Mucinous	Others
Rosen [15]	75 % Ductal	10 %	10 %	1 %	2 %	---
Fisher [16]	58 % NOS	5 %	6 %	1 %	2 %	28 % Combined
Page [17]	70 % NST	10 %	5 %	3 %	2 %	2 % Combined

NST-no special type;   NOS – not otherwise specified

- The incidence of special types of invasive breast cancers is approximately 20- 30 % of all invasive breast cancers. The remaining 70 – 80 % is invasive "ductal" or no special type (NST), "a term of exclusion that reinforces the practice of recognizing and reporting special types of invasive carcinoma. Special types of carcinoma are defined by recognizable histologic features present homogenously throughout the lesion (>90 % of the lesion). When special features are present in 75 % to 90 % of the carcinoma (variant) an improved prognosis is recognized.

Table – 1: Special types of invasive ductal carcinoma
I.   Tumors with a better prognosis
        A. Tubular carcinoma
        B. Mucinous carcinoma
        C. Medullary carcinoma
        D. Invasive cribriform carcinoma
        E. Adenoid cystic carcinoma
        F. Invasive papillary carcinoma

II.   Other special type of invasive ductal carcinoma
         A. Invasive micropapillary carcinoma
         B. Apocrine carcinoma
         C. Neuroendocrine carcinoma
         D. Metaplastic carcinoma
         E. Other rare subtypes.

- Tubular carcinoma is probably the most important special type of carcinoma because of its relatively high incidence and certainty of prediction. [18] When present in pure form, it is highly unlikely to have distant metastatic potential. Invasive cribriform carcinoma is very similer, both histologically and biologically to tubular carcinoma. [19] Another excellent prognosis carcinoma is mucinous (colloid), defined by the presence of pools of extracellular mucin within which aggregates of low grade tumor cells appear to float. [20]

- Generally, special type carcinomas are of low grade. An exception is medullary carcinoma, characterized by microscopic features of a high grade malignancy: lack of tubule formation, nuclear pleomorphism, and high mitotic rate. Its other defining histologic elements indicate a much better prognosis than that indicated by grading alone.

I.   Tumors with a better prognosis

A. Tubular carcinoma

- Tubular carcinoma is a type of invasive ductal carcinoma with excellent prognosis, comprising of well defined tubular structures with open lumina lined by single layer of epithelial cells with a loose cellular fibrous stroma. The pure form of tubular carcinoma accounts for approximately 2 % of invasive ductal carcinoma; the frequency is higher in series with only T1 tumors or in series with screening mammography detected breast cancers. [21- 24] In comparison to the invasive ductal carcinoma NOS, these tumors tend to occur in older age group, are smaller in size and have significantly less nodal metastasis.

Pathology:

- Tubular carcinoma range in size from 0.2-2.0 cm, majority being less than 1 cm. Grossly tow morphologic subtypes have been described with include the pure type with a pronounced stellate configuration and radiating arms and the sclerosing subtype characterized by an ill defined diffuse structure. [21 - 25] On histology characteristic feature of tubular carcinoma is the presence of open tubular structures within a loose cellular stroma. The tubules are often angulated and lined single layer of bland epithelial cells; some limning cells may contain apocrine snouts. Most people believe that in the pure form of tubular carcinoma, these tubular structures must comprise >90 % of the infiltrative carcinoma. Tumors with these features in 75-90 % of tumor must be considered as variants of tubular carcinoma and classified as grade I invasive ductal carcinoma, often referred to as mixed tubular carcinoma. [29] Tubular carcinoma is usually estrogen and progesterone receptor positive and has a low proligerative index. Tubular carcinoma is often associated with epithelial proliferative lesions such as low grade ductal carcinoma in situ (DCIS), cribriform and micropapillary type in majority of the lesions. Other proliferative lesions seen in association with tubular carcinoma include intra ductal and atypical ductal hyperplasia (ADH) including columnar alteration with prominent apical snouts and secretions (CAPSS), atypical lobular hyperplasia, lobular carcinoma in situ, and radial scar. [21,26-28]

Differential diagnosis:

- **Sclerosing adenosis:** is distinguished by its characteristic lobulo-centric architecture and presence of myoepithelial cells in which Immunohistochemistry using myoepithelial markers (calponin, smooth muscle actin, p63 and CD10) can be very helpful.

30

- **Microglandular adenosis (MA):** the tubules of MA while may lack myoepithelial cells are more rounded and regular and contain eosinophilic colloid like material in the lumen; also a ring of basement membrane is present around the tubules.

- **Radial scar/complex sclerosing lesion:** RS has a characteristic radial architecture and myoepithelial cells can be identified in the tubular structures.

Prognosis:
- Pure tubular carcinoma has an excellent prognosis, which in some series is similar to age matched women without breast cancer. [21] Recurrence following breast conserving surgery or mastectomy is rare and axillary node metastasis occur infrequently, in approximately 9 % cases and may be higher in mixed tubular carcinoma. [14]

- The importance of the diagnosis of tubular carcinoma is identifying an excellent subset of invasive ductal carcinoma, which may have significant therapeutic implications. Some people have suggested that pure tubular carcinoma <1 cm in size may be treated with excision alone. [29, 30]

B. Mucinous or colloid carcinoma

- This is another special type of breast cancer with excellent prognosis, which is characterized by islands of small and uniform cells floating in large amounts of mucin. Pure mucinous carcinoma, which includes tumors with more than 90 % component being mucinous, accounts for approximately 2 % of all breast cancers. It occurs in all ages but the mean age is higher than the invasive breast carcinoma NOS.

Pathology:
- The size of pure mucinous carcinoma range from 1 to 20 cm, with an average size of less than 5 cm in majority of cases. [14] In the pure form of mucinous carcinoma the consistency is gelatinous and soft. On histology

lakes of extra cellar mucin are seen with islands of small uniform tumor cells. Fibrous septa may intersect the lakes of mucin. Often solid and micropapillary 6 type DCIS is seen associated. Two variants of mucinous carcinoma have been described, pure (with >90 % mucinous areas) and mixed variants. Significant proportion of these tumors may show neuroendocrine differentiation with chromogranin and synaptophysin immunoreactivity. [21]

Differential diagnosis:
- **Mucocele like lesion (MLL):** In MLL, presence of myoepithelial cells adhering to the strips of epithelial cells floating in the mucin is helpful in differentiating from mucinous carcinoma. Also dilated mucin filled ducts may be seen in the vicinity.

- **Myxoid fibroadenoma:** The presence of compressed spaces lined by epithelial and myoepithelial cells within a myxoid stroma with mast cells are helpful in the diagnosis of myxoid fibroadenoma.

Prognosis:
- Pure form of mucinous carcinoma has an excellent prognosis with 10 years survival ranging from 80 to 100 %. The prognosis is worse in mixed tumors and depends on the grade of the non-mucinous component. Incidence of lymph node metastasis in pure mucinous carcinoma is 3-15 % compared to 33-46 % seen in mixed type. The presence or absence of neuroendocrine features does not affect the prognosis.

C. Medullary carcinoma (MC)

- This rare form of invasive ductal carcinoma is characterized by a well circumscribed tumor with soft consistence composed of poorly differentiated cells arranged in large sheets, with no glandular structures, and scant stroma with abundant lymphoid infiltrate. The incidence of MC ranges from 1-7 % depending on the stringency of the diagnostic criteria used. [21] In its pure form it is extremely rare and most people agree that very strict criteria must be used in the diagnosis of pure MC because of the good prognosis associated with it and also significant therapeutic implications that may follow after such a diagnosis.

Pathology:
- Grossly MC is rounded, well- circumscribed tumor with well-defined margins, soft consistency and fleshy tan to gray cut surface. The median diameter ranges from 2-3 cm, which is similar to the usual type of ductal carcinoma. [14, 21]

On histological examination, classically five features characterize MC:
1. Complete histological circumscription with pushing margin.
2. A syncitial architecture (tumor cells arranged in sheets usually 4-5 cell thick) which should be observed in >75 % of the tumor.
3. Absence of glandular and tubular structures.
4. Diffuse and prominent lymphoplasmacytic infiltrate.
5. Tumor cells are large, rounded with abundant cytoplasm, vesicular nuclei containing prominent nucleoli. There is marked nuclear pleomorphism with numerous mitoses and atypical giant cells may be seen.

- Tumors in which some but not all the diagnostic criteria are present have been referred to as atypical medullary carcinoma (AMC). Some people have suggested not using the AMC category and infiltrating ductal carcinoma with medullary features may be the most appropriate terminology for such tumors. [21]

- Flow cytometry and immunohistochemical analysis have shown that most medullary carcinoma are aneuploid and highly proliferative lesions. They typically lack estrogen receptor expression and have low incidence of HER-2 overexpression. A high frequency of MC has also been reported in patients with BRCA1 germ line mutation. [21]

Prognosis:

- MC has been reported to have a better prognosis than the usual type of infiltrating ductal carcinoma, but has been questioned by other studies. [21, 31] The overall 10 year survival reported for MC varies between 50 % to more than 90 %. This difference in survival may be due the difference in diagnostic criteria for MC. [21]

D. Invasive cribriform carcinoma (ICC)

- Invasive cribriform carcinoma is a well differentiated variant of ductal carcinoma with excellent prognosis that grows in a largely cribriform pattern, often mixed with a tubular carcinoma component; these are termed classic cribriform carcinoma. It resembles tubular carcinoma both histologically and biologically. As with tubular carcinoma, purity of pattern is essential to ensure excellent prognosis. ICC accounts for 0.8-3.5 % of breast carcinoma and the mean age of the patients is 53-58 years. Multifocality is observed in 20 % of cases. [14, 21, 29]

Pathology:

- The pure ICC consists almost entirely (>90 %) of an invasive cribirform pattern. Apical snouts are a regular feature; the tumor cells are small and show very little pleomorphism. Cribriform type DCIS is seen almost 80 % cases. Axillary lymph node metastasis occurs in approximately 14 % cases and the cribriform pattern is preserved in the nodal metastasis. Tumor showing predominantly cribriform patter with a minor (<50 %) component of tubular carcinoma are also included in this category of classic ICC. The term mixed invasive cribriform carcinoma is reserved for tumors in which <50 % of the tumor is composed another carcinoma type, other than tubular carcinoma (WHO). If a higher-grade area of 3-5 mm is present, the excellent prognosis associated with ICC is not guaranteed. [29,32]

Differential diagnosis:

- Carcinoid tumor is distinguished by its characteristic nueroendocrine differentatiation on Immunohistochemistry.

- Adenoid cystic carcinoma is shows second (myoepithelial) cell type, and intracystic secretory and basement membrane like material.

- Cribriform type DCIS is distinguished from ICC by presence of myoepithelial cells around the nests of tumor cells.

Prognosis:
- ICC has an excellent survival with 10 year survival being 91 % [31], to 100 %. [32]

E. Adenoid cystic carcinoma

- Adenoid cystic carcinoma (ACC) is a special type of invasive ductal carcinoma with low aggressive potential histologically similar to the salivary gland counterpart. ACC represents about 1 % of breast carcinoma and age distribution is similar to the usual type of ductal carcinoma. About 50 % of ACC are found in the sub-perialveolar region, they usually present as a discrete nodule, which may sometimes be painful. [21]

Pathology:
- The size of ACC varies from 0.7 to 12 cm with an average of 3 cm in most series. On gross examination tumors are usually circumscribed an may have microcysts. On histology ACC is similar its alivary gland counterpart comprising of two cell types, the basaloid cells and the inner glandular cell type; with three basic patterns: trabecular-tubular, cribriform and solid. Adenomyoepitheliomatous and syringomatous areas are also present in some tumors and sebaceous differentiation is seen 14 % caes suggesting the structural diversity of this tumor and possible histogenetic association with epiehtlial-myoepithelial type tumors. Estrogen receptor is negative in virtually all ACC. [21, 33,34] Some series have reported association of ACC with microglandular adenosis (MGA) suggesting transition from MGA to ACC. [35]

Immunohistochemistry:

- On Immunohistochemistry basaloid cells of ACC are positive with vimentin, CK14, and focally for myoepithelial markers such as smooth muscle actin, calponin, p63 and maspin; ultrastructurally basaloid cells show myoepithelial features, such as thin Cytoplasmic filaments and well-developed desmosomes. The glandular cell type is usually positive for CK7. [36] The basaloid cells are present at the periphery and because of their polarity they express laminnin, fibronectin, basal lamina related proteins, and type IV collagen' while the glandular luminal cells express proteins related to cell polarization and epithelial differentiation including E-cadherin, and beta catenin. [36]

Differential diagnosis:

- Collagenous spherulosis described by Clement et al in 1987 [37] is a benign duct proliferation, which combines gland formation and acellular deposits (spherules) of stromal material among the epithelial cells, this can sometimes mimic ACC and should be distinguished by the presence of adjacent bening proliferative disease. In some cases collagenous spherulosis have been found associated with ACC, which can pose a problem especially in needle core biopsy. [38]

- Cribriform type DCIS is distinguished from ACC by showing only on cell type of lesion and lack of basement membrane like material. Also virtually all cribriform DCIS are ER positive while ACC is typically ER negative.

Prognosis:

- ACC is a low-grade malignant tumor with mastectomy being curative in most cases reported. It rarely spreads through lymphatics; local recurrence may be related to incomplete resection. Distant metastasis occurs in about 10 % cases and lung is the most common site.

F. Invasive papillary carcinoma

- Invasive papillary carcinoma (IPC) is the invasive carcinoma arising in the association with intracystic papillary carcinoma and having a papillary growth pattern in the invasive component. They comprise 1-2 % of invasive breast cancer and are characterized by a relatively good prognosis. IPC is diagnosed mostly in postmenopausal women; mammography shows a rounded, well-circumscribed density with a radiological differential diagnosis of fibroadenoma, medullary and mucious carcinoma.

Pathology:
- On gross examination most IPC are well-circumscribed tumors. On histology they are circumscribed tumors with delicate or blunt papillae infiltrating the stroma often with some solid areas. DCIS is present in 75 % cases and usually but not always is of papillary type. Estrogen receptor is usually positive. [21,33]

Prognosis:
- The data on prognosis in IPC is limited and the overall survival usually depends on the stage of the tumor. However in node negative tumors, patients with IPC histology showed an improved 10 years survival compared to invasive ductal carcinoma NOS. [40]

II.   Other special types of invasive ductal carcinoma

A. Invasive micropapillary carcinoma

- Invasive micropapillary carcinoma (IMC) is a tumor composed of small clusters of tumor cells lying within clear stromal spaces resembling dilated lymphatic channels. Carcinoma with dominant mircopapillary growth pattern account for less than 2 % of all invasive breast cancers. Foci of micropapillary pattern may however be seen in up to 6 % of invasive breast cancer. These tumors tend to show a higher frequency of lymph node metastasis, which may be evident at the time of presentation.

Pathology:
- On gross examination IMC has lobulated outline due to its expansive mode of growth. On histology it consists of h aggregates of malignant cells, which on cross section have an appearance of a tubule with obliterated lumen lying in artifactual stromal spaces that mimic lymphatics but lack lining endothelial cells. In non-pure forms of this tumor transition from usual type ductal carcinoma to IPC may be seen. Peritumoral lymphovascular invasion is fairly common seen in up to 60 % cases. [21] In lymph node and pleural metastases the micropapillary pattern is preserved.

Differential diagnosis:
- Metastatic serous carcinoma of the ovary to the breast: Rare cases of metastasis from an ovarian primary to axillary lymph nodes and breast have been reported which may be confused with primary IMC of the breast. Immunohistochemistry can be helpful in this situation with WT1 immunostaining positive in ovarian primary and Gross Cystic disease Fluid Protein (GCDFP-15) and mammaglobin staining seen in breast primary. [41]

Prognosis:
- Incidence of axillary node metastasis is more common in IMC. However, in multivariate analyses a micropapillary growth pattern has no independent significance for survival. [21]

B. Apocrine carcinoma

- Apocrine carcinoma (AC) is defined as invasive carcinoma showing predominant (>90 %) cytological and immunohistochemical features of apocrine cells. The incidence of AC ranges from 0.4-4 %. There is no difference in clinical and radiological features of AC from usual type ductal carcinoma.

Pathology:

- Apocrine cells can be seen in most type of breast carcinoma. Predominant apocrine features may be seen as two cell types. Type A apocrine cells composed of large cells with ampler eosinophilic finely granular cytoplasm. The granules are PAS positive after diastase digestion. The nuclei vary from globoid with prominent nucleoli to hyperchromatic. Tumors with pure type A cells mimic granular cell tumor and have been referred to as myoblastomatoid. [42] type B cells have a foamy cytoplasm mimicking foamy histiocytes; they have also been designated as sebocrine cells.

Differential diagnosis:

- Immunohistochemistry using cytokeratin antibody can be very useful in differentiating AC from foamy histiocytes and granular cell tumor.

Prognosis:

- There are no differences in survival in AC compared to the usual type of ductal carcinoma.

C. Neuroendocrine carcinoma (NEC)

- Primary neuroendocrine carcinoma (NEC) of the breast include tumors in which more than 50 % of the tumor shows exhibits morphological and immunohistochemical of neuroendocrine differentiation as seen in the gastrointestinal tract and lung. Focal neuroendocrine differentiation on Immunohistochemistry may be seen usual type of ductal carcinoma; they are not included in the group. NEC represents 2 - 5 % of all breast cancers. Most patients are in 6th or 7th decades of life; there are no notable differences in presentation from other tumor types. Patients often present with palpable nodule, which usually appears circumscribed on mammographic and ultrasound examination.

Pathology:

- On gross examination NeC may appear as infiltrative or expansile tumors. On histology, most tumors form alveolar structures or solid sheets of cells with a tendency to produce peripheral pallisading. Tumors may present as different cell types:

1. **Solid neuroendocrine carcinoma:** These tumors consist of densely cellular, solid nests and trabeculae of cells that vary from spindle to plasmacytoid and large clear cells [43] separated by delicate fibrovascular stroma. The tumor cells may form well-defined lobulated masses. Some of these tumors appear to originate from solitary, solid papillary intraducatal carcinoma. Others form multiple, often rounded solid nests.
2. **Small cell/Oat cell carcinoma:** This is morphologically similar to the small cell carcinoma of the lung composed of densely packed hyperchromatic cells with scant cytoplasm, high mitotic index and often associated with necrosis.
3. **Large cell neuroendocrine carcinoma:** These are composed of crowded large clusters of cells, with moderate to abundant cytoplasm, nuclei with vesicular to finely granular chromatin and high mitotic index. These are also similar to their lung counterpart.

Differential diagnosis:
- Metastatic neuroendocrine carcinoma: Immunohistochemistry may be helpful using antibodies to estrogen receptor GCDF. In addition presence of solid type DCIS supports a primary breast carcinoma.

- Infiltrating lobular carcinoma may be differentiated from small cell carcinoma by e-cadherin immunostaining, which is positive in lobular carcinoma.

Prognosis:
- The prognosis of NEC including small cell carcinoma of the breast depends on the stage of the disease. [21]

D. Metaplastic carcinoma

- Mammary carcinoma may undergo metaplasia to another epithelial cell type such as squamous phenotype or into a mesenchymal phenotype such as spindle cells or other Heterologous elements including chondroid, myxoid, osseous or lipomatous. In some cases the metaplastic component may completely replace the normal glandular component and no recognizable adenocarcinoma may be identified. Clinical presentation of metaplastic carcinoma is not different from the usual type of ductal carcinoma. On mammography the tumors tend to have well-circumscribed contours. With few exceptions metaplastic carcinoma is estrogen receptor negative.

Pathology:
- The reported size of metaplastic carcinoma ranges from 1 to 21 cm, with a mean or median size of 3 to 4 cm. The tumors are typically described as solid, firm to hard, nodular and circumscribed mass. On histology metaplastic carcinoma can be divided into broad subtypes depending on the phenotypic appearance of the tumor. These include:

1. **Squamous cell carcinoma:** The most common metaplastic pattern in mammary carcinoma is focal squamous metaplasia in an otherwise typical ductal adenocarcinoma. The spectrum of squamous differentiation ranges from mature keratinizing epithelium to poorly differentiated carcinoma with spindle cell pseudosarcomatous areas. In spindle cell variant virtually the entire tumor is replaced by pseudosarcomatous growth pattern [2, 3], therefore extensive sampling is essential to identify squamous or adenocarcinoma component.
2. **Adenocarcinoma with spindle cell metaplasia:** In this form of metaplastic carcinoma, adenocarcinoma is seen admixed with abundant spindle cell component which are neither squamous cells or mesenchymal, but are thought to be of glandular phenotype. The spindle cells are immunoreactive with epithelial markers such as cytokeratin 7, but not CK5/6 or other markers of squamous and myoepithelial differentiation. Ultrastructurally the spindle cells contain intracytoplasmic lumens confirming glandular cell population.
3. **Adenosquamous carcinoma:** In this form of metaplastic carcinoma, glandular and squamous components are seen intermixed.

4. **Low grade adenosquamous carcinoma:** This is a variant of metaplastic mammary carcinoma with characteristic histologic appearances. It is morphologically similar to adenosquamous carcinoma of the skin and has been referred to by some as syringomatous squamous tumor or infiltrating syringomatous adenoma. On histology these tumors are composed of small glandular structures and solid cords of epithelial cells haphazardly arranged in an infiltrative spindle cell stromal component. The proportion of these components may vary. They have a tendency to grow around and in between ducts and lobules especially at the periphery of the lesion. Squamous metaplasia varies from syringomatous like appearance to inconspicuous foci in largely glandular lesions. Small osteocartilagenous foci have been reported in some cases. The majority of cases have an excellent prognosis, but a subset can behave in a locally aggressive manner; recurrence is related to the adequacy of excision.

5. **Mixed epithelial/ mesenchymal metaplastic carcinoma:** This has been referred to as carcinoma with chondro-osseous metaplasia or matrix producing carcinoma. On histology infiltrating ductal carcinoma is seen intermixed with malignant spindle cells and Heterologous mesenchymal elements ranging from bland chondroid and osseous differentiation to frank sarcoma (chondrosarcoma, osteosarcoma, liposarcoma, rhabdomyosarcoma, fibrosarcoma). The spindle cell component may show positive immunoreactivity to cytokeratin. Chondroid component may co-express S100 and cytokeratin.

Differential diagnosis:
- In tumors lacking significant epithelial component, the main differential diagnosis is Primary breast sarcoma and myoepithelial carcinoma (malignant myoepithelioma). Immunohistochemistry using epithelial markers such as cytokeratin, EMA may be helpful in identifying the epithelial lineage and establishing a diagnosis of metaplastic carcinoma. In some cases extensive sampling is needed to identify a ductal carcinoma component or in some cases the presence of DCIS may be useful.

Prognosis:

- The incidence of axillary node metastasis is less in metaplastic carcinoma compared to the same size of usual ductal carcinoma; 10 – 15 % of pure squamous cell carcinoma and 19 – 25 % of chondro-osseous type metaplastic carcinoma show axillary node metastasis. In limited series the 5 year survival in metaplastic carcinoma ranges from 28 – 68 %. As expected advanced stage and lymph node metastasis is associated with more aggressive behavior.

E. Other subtypes

- Some extremely rare histologic types of invasive ductal carcinoma include Glycogen-rich, clear cell carcinoma (GRCC), sebaceous carcinoma, acinic cell carcinoma, oncocytic carcinoma, secretory carcinoma, lipid-rich carcinoma, signet ring cell carcinoma (non-lobular type, mucinous cystadenocarcinoma and columnar cell mucinous carcinoma. [21]

- Histological typing and grading of invasive ductal carcinoma has been the standard of care over the years and still remains one of the most important factor in the evaluation of breast cancer to predict behaviour and plan further management of these cases. Advances are being made in the field of genomics and proteomics and it is possible that in the future these newer molecular diagnostic tools may be better able to characterize breast cancer in various subgroups according to their behaviour and response to different types of adjuvant therapy avoiding some degree of subjectivity involved in the histologic evaluation. [44] Some recent studies have classified breast cancer into subgroups with clinical implications on the basis of gene expression patterns. **Sorlie et al.** [45] classified tumors into basal epithelia like subgroup, an HER-2-overexpressing sub group and a normal breast (luminal cell) like group. Other studies have also confirmed these findings and immunohistochemical studies have shown a specific cytokeratin phenotype in which the breast luminal type tumors immunoreact with cytokeratin 7/8, 18, 19 and the basal like carcinoma of the breast react with CK5/6. The breast luminal cell like carcinoma are estrogen receptor positive and have a better prognosis as opposed to basal like carcinoma which are usually estrogen receptor negative and have a worse prognosis. [46]

- In addition to identifying clinically relevant subgroups such as basal like carcinoma some recent studies have identified subsets of genes that are unregulated and down regulated in tumors with differing biological behavior. [47] In a recent study using the NSABP database Paik et al have analyzed a set of 21 genes including 5 housekeeping genes to develop an assay (referred to as Oncotype) which can help predict outcome and response to therapy in tamoxifen treated nod negative breast cancer. [48] As our understanding of these newer techniques improves hopefully we will be better able to apply them in the evaluation and management of breast cancer in routine practice.

PAGET'S DISEASE OF THE NIPPLE:
- Background of keratin, squamous cells and inflammatory cells (scrape smears from nipple).
- Large malignant cells, single, and in small groups.
- Abundant pale cytoplasm with distinct borders.
- Obvious nuclear features of malignancy.

- **HISTOLOGY:** Large clear cells with atypical nuclei are seen within the epidermis, usually concentrated along the basal layer but also permeating the malphigian layer.

INFILTRATING LOBULAR CARCINOMA:
- A variable, often very poor cell yield.
- Cells single and in small clusters short single files common.
- Cytoplasm scant, indistinct, often missing.
- Small dark nuclei of relatively uniform size.

- **HISTOLOGY:** Presence of small and relatively uniform tumor cells growing singly, in indian file pattern, and in a concentric fashion around lobules involved by in situ lobular hyperplasia. The stroma is usually abundant, is of dense fibrous type, and contains foci of periductal and perivenous elastosis.

- Variants are histiocytoid, pleomorphic, signet ring type, tubulo-lobular, and alveolar.

ADENOMYOEPITHELIOMA:
- **HISTOLOGY:** Balanced proliferation of round, oval or tubular glandular elements with intervening islands and bands of polygonal myoepithelial cells that have clear cytoplasm.

PRIMARY STROMAL SARCOMA:
- A diagnosis of mammary sarcoma can be established only after metaplastic carcinoma has been excluded. The distinction is important for treatment as well as for prognosis. The lesion should be sampled extensively for evidence of in situ or invasive carcinoma.

REFERENCES

1. Rosai Juan. Breast. In: Rosai and Ackerman's Surgical Pathology. Mosby; 2004, p.1763-1876.
2. Srinivas M, Kumar H G, Reddy J S, Bhakaran SC. Role of Fine Needle Aspiration Cytology in the Diagnosis of Breast Lumps and it's Histopathological Correlation. Indian J. Pathol. Microbiol. 1989; 32: 2; 133 – 137.
3. WHO classification of tumors 2003, Pathology and genetics of tumors of breast and female genital organs.
4. Orell S, Sterret G, Whitaker D, Lindholm K. Breast. Fine needle Aspiration Cytology. 4th edition. Churchill Livingstone; 2005, p.165-225.
5. Parkin DM, Bray F, Ferlay J, et al. Estimating the world cancer burden: Globocan 2000, Int J Cancer, 2001; 94: 153-156.
6. Swanson G M, Lin C S. Survival patterns among younger women with breast cancer: the effects of age, race, stage, and treatment. J Natl CancerInst Monogr.1994; 16: 69-77.
7. Bertheau P, Steinberg S M, Merino M J. cerb B2, p53, and nm23 gene product expression in breast cancer in young women: Immunohistochemical analysis and clinicopathologic correlation. Hum Patholgy. 1998; 29: 323-9.
8. Largent JA, Ziogas A, Anton-Culver H. Effect of reproductive factors on stage, grade and hormone receptor status in early onset breast cancer. Breast Cancer Res. 2005; 7: 541-54.
9. Albain K S, Allred DC, Clark GM. Breast cancer outcome and predictors of outcome: Are there age differentials? J Natl Cancer Inst Monogr. 1994; 16: 35-42.
10. Nixon A J, Neubreg D, Hayes D F, et al. Relationship of patient age to pathologic features of the tumor and the prognosis for patients with stage I or II breast cancer. J Clin Oncol. 1994; 12: 888-94.
11. Bonnier P, Romain S, Charpin C, et al. Age as a prognostic factor in breast cancer: Relationship to pathologic and biologic features. Int J Cancer. 1995; 62: 138-44.

12.Zabicki K, Colbert JA, Domingyez FJ, et al. Breast cancer Diagnosis in Women <40 versus 50 to 60 years: Increasing size and stage disparity compared with older women over time. Annals of Surgical Oncology. 2006 Aug; 13(8):1072-1077.

13.Fernandopulle S, Cher-Siang Peter, Hoon Tan P, et al. Breast carcinoma in women 35 years and younger: A pathological study. Pathology J. 2006 June; 38(3): 219-222.

14.Rosen PP, Obermari HA, AFIP Atlas of Tumor Pathology: Tumors of the Mammary Gland. 3rd series, AFIP Press, 1993.

15.Rosen PP. The pathological classification of human mammary carcinoma: Past, present, future. Ann Clin Lab Sci. 1979; 9: 144-156.

16.Fisher ER, Gregorio RM, Fisher B: The pathology of invasive breast cancer. A syllabus derived from findings of the national surgical adjuvant breast project (protocol no. 4).Cancer. 1975; 36: 1-85.

17.Page D L, Anderson T J, Sakamato G: Infiltrating carcinoma: Major histological types, in page DL, Anderson TJ (eds): Diagnostic histology of the breast. New York, NY, Churchill Livingstone, 1987, p. 193-235.

18.Simpson JF, Page DL. Prognostic value of histopathology in the breast. Semin Oncol. 1992 June; 19(3): 254-262.

19.Venable JG, Schwartz AM. Silverberg SG: Infiltrating cribriform carcinoma of the breast: A distinctive clinicopathologic entity. Hum pathol. 1990 Mar; 21(3): 333-338.

20.Komari K, Sakamato G, Sugano H H, et al. Mucinous carcinoma of the breast in Japan: A prognostic analysis based on morphologic features. Cancer. 1988; 61: 989-996.

21.World health organization classification of tumors, pathology and genetic: Tumors of the breast and female genital organs, Ed. Tavassoli FA, Devilee P, IARC Press, Lyon France, 2003.

22.Meyer JS. Cell kinetics of histologic variants of in situ breast carcinoma. Breast Cancer Res Treat, 1986; 7: 171-180.

23.Patchefsky AS, Shaber Gs, Schwartz GF et al. The pathology of breast cancer detected by mass population screening, Cancer, 1977; 40: 1659-1670.

24. Rajakariar R, Walker RA. Pathological and biological features of mammographically detected invasive breast carcinomas. Br J Cancer, 1995; 71: 150-154.

25. Ellis IO, Galea MH, Locker A, et al. Early experience in breast cancer screening: Emphasis on development of protocols for triple ssessment. Breast, 1993; 2: 148-153.

26. Sloane JP, Mayers MM. Carcinoma and atypical hyperplasia in radial scars and complex sclerosing lesions: Importance of lesion size and patient age. Histopathology, 1993; 23: 225-231.

27. Fraser JL, Raza S, chorny K, et al. Columnar alteration with prominent apical snouts and secretions: A spectrum of changes frequently present in breast biopsies performed for microcalcification. Am J Surg Pathol, 1991; 22: 1521-1527.

28. Goldstein NS, O'Malley BA. Cancerization of small ecstatic ducts of the breast by ductal carcinoma in situ cels with apocrine snouts: A lesion associated with tubular carcinoma. Am J Clin Pathol. 1997; 107: 561-566.

29. Page DL. Special types on invasive breast cancer, with clinical implications. Am J Surg Pathol., 2003; 27: 832-835.

30. Baker RR. Unusual lesions and their management. Surg Clin North Am., 1990; 70: 963-975.

31. Ellis IO, Galea M, Broughton N. Pathologic prognostic factors in breast cancer. II. Histologic type. Relationship with survival in a large study with long term follow-up. Histopathology, 1992; 20: 479-489.

32. Page DL, Dupont WD. Anatomic markers of human premalignancy and risk of breast cancer. Cancer, 1990; 66: 1326-1335.

33. Shin SJ, Rosen PP. Solid variant of mammary adenoid cystic carcinoma with basaloid features: A study of nine cases. Am J Surg Pathol. 2002; 26: 413-420.

34. Bennet AK, Mills SE, Wick MR. Salivary type neoplasms of the breast and lung, Semin Diagn Pathol., 2003; 20: 279-304.

35. Acs G, Simpson JF, Bleiweiss, et al. Microglandular adenosis with transition into adenoid cystic carcinoma of the breast. Am J Surg Pathol., 2003; 27: 1052-60.

36. Pia-Foschini M, Reis-Filho JS, Eusebi V, Lakhani SR. Salivary gland like tumors of the breast: Surgical and molecular pathology. J Clin Pathol, 2003; 56: 497-506.

37. Clement PB, Young RH, Azzopardi JG. Collagenous spherulosis of the breast. Am J Surg Pathol., 1987; 11: 411-417.

38. Ogata K, Sakamoto G, Sakuri T. Adenoid cystic carcinoma with collagenous spherulosis like structures in the breast: Report of a case. Pathology International, 2004; 54: 332-336.

39. Rosen's Breast Pathology. Ed Rosen PP, Lippincott Raven, 1997.

40. Fisher Er, Anderson ER, Redmond C, et al. Pathologic findings from the National Surgical Adjuvant Breast Project protocol B-06. 10 year pathological and clinical prognostic discriminants. Cancer, 1993; 71: 2507-2514.

41. Recine MA, Deavers MT, Middleton LP. Serous carcinoma of the ovary and peritoneum with metastases to the breast and axillary lymph nodes. A potential pitfall. Am J Surg Pathol., 2004; 28: 1646-1651.

42. Eusebi V, Foschini MP, Bussolati G, et al. Myoblastomatoid (histiocytoid) carcinoma of the breast. A type of apocrine carcinoma. Am J Surg Pathol., 1995; 19: 553-562.

43. Sapino A, Righi L, Cassoni P, et al. Expression of neuroendocrine phenotype in carcinomas of the breast. Semin Diagn Pathol., 2000; 17: 127-137.

44. Perou CM, Sorlie T, Eisen MB, et al. Molecular portraits of human breast tumors. Nature, 2000; 406: 747-752.

45. Sorlie T, Perou CM, Tibshirani R, et al. Gene expression patterns of breast carcinomas distinguish tumor subclasses with clinical implication. PNAS 2001; 98: 10869-10874.

46. El-Rehim A, Pinder SE, Paish CE, et al. Expression of luminal and basal cytokeratins in human breast carcinoma. J Pathol., 2004; 203: 661-671.

47. Van De Vijver MJ, He YD, Van T Veer LJ, et al. A gene expression signature as a predictor of survival in breast cancer. NEJM 2002; 347: 1999-2009.

48. Paik S, Shak S, Tang G, et al. A multigene assay to predict recurrence of tamoxifen treated nod negative breast cancer. NEJM 2004; 351: 2817-2826.

www.ingramcontent.com/pod-product-compliance
Lightning Source LLC
Chambersburg PA
CBHW021934170526
45157CB00005B/2312